Robotics and

Automation

Industry

Thought

Leaders 2

Robotics and Automation Industry Thought Leaders 2

This is the second edition or volume of this book of interviews from the Robotics and Automation News website. Since our launch last year, we have enjoyed a moderate amount of success, with an increasing number of people visiting the site and signing up to our free newsletter.

However, in order to continue our operations, we need revenue. We have created a few streams through which readers can contribute.

Firstly, we have set up what's called a paywall. At the moment, our paywall is voluntary, in that you can just close the popup advertisement which asks you to sign up and continue reading. We feel this is reasonable since we obviously do not want to lose readers, and we recognise the fact that some may not be able to afford to pay. Most of us have been there.

However, for those of you who are able to pay, we would appreciate your support, which you can show by subscribing to the website next time you see the popup. That way, you don't see that popup any more.

Another channel we have opened in the hope of generating some income is book publishing. This is the second book of interviews from the website. We feel it provides insight from industry professionals who we consider to be thought leaders in their respective fields.

More books are planned, but we hope you will understand that without the necessary resources, it is difficult to produce books, which can take a long time to research, write and prepare for market. So, here, too, we need your support.

The fact that you are reading this page means that you are one of our supporters, and we very much appreciate your investment in us.

Robotics and Automation News has a real chance to become the most important resource for this industry. When we look around the web, we get the sense that there is nothing like our website and the information we are offering. Some websites include the subject as part of a larger information package, but no one deals with it specifically like we do.

This encourages us to think we can become an essential outlet for the industry, and we look forward to investing in more in-depth interviews and features which we hope you will find valuable in your work and business, as well as of general interest to you.

Thank you. ●

Chapter 2 : Erik Walenza-Slabe

The IoT with Chinese characteristics

Interview with Erik Walenza-Slabe, CEO of IoT One

Apparently, we are all either living in the Age of Industry 4.0 already, or we are entering it. And one of the main features of this age is a relatively new connectivity technology called the "internet of things", often abbreviated to IoT.

The IoT is, as you might imagine or know, a network of "things". These things can be computers, industrial robots or any other type of robot, any device or appliance – anything that is a thing or a machine, which is why it's also sometimes referred to as the machine-to-machine network.

Perhaps it's not a particularly accurate label for it, mainly because the IoT can be part of the general internet that we all know and use, in that it can use the same Ethernet cables and WiFi and whatever other connections there are available, although it's thought M2M communications could mainly be carried through radio-frequency identification technology.

However, any connection system capable of carrying internet protocol data could also carry M2M communications, and therefore be considered part of the IoT. So maybe there will not be much of a distinction going forward. Perhaps it might help to think of the IoT like the worldwide web for machines.

And while it might sound like a gimmick to some, the IoT is actually very serious business. Cisco estimates that around 50 billion machines will be on the IoT by the end of this decade. Some estimate say the number will be even higher.

New and very large radio networks are being built specifically for the things. So, in the future, your fridge can tell the smart trolley at your local grocery store that you've run out of TV dinners, and the smart trolley can locate that vegetarian meal you like and load it onto the driverless delivery van, which can bring it to your home robot, which can make the payment and put your food in the fridge, maybe even warm it up and place it in front of the TV just in time for when you get home after work. In the future, you might have no idea that you own a fridge.

The IoT looks like it will initially have most impact on industry. The high levels of automation within sectors such as manufacturing and logistics seem to have been designed with what is being called the "industrial internet of things", or IIoT, which some see as the foundation of Industry 4.0.

General Motors has recently implemented a networking solution which connects all the industrial robots at work in their factories around the world. Although the intention is to enable a centralised group of humans in some GM control control somewhere to monitor the machines, the obvious next stage is to allow the robots to communicate with each other.

So, instead of the robot telling the control room it needs a drink of oil otherwise it's going to feint from the heat, it can just tell the nearest mobile vending machine to dispense its favourite beverage.

The IoT has absolutely massive global potential for accelerating industries such as manufacturing and logistics. Many countries are hoping to effectively implement IoT and other Industry 4.0 technologies to accelerate their manufacturing growth, one of them being Germany, where they call it "Industrie" 4.0. However, right now, when you think of countries associated with manufacturing and logistics, you might think of China first.

Estimated to have overtaken the US as the single largest manufacturing nation, it is increasingly interested in robotics and automation technologies, which also will inevitably account for much of the data traffic on the thingyverse's information superhighway.

So who better to learn more about the subject than from someone involved in IoT in China?

Erik Walenza-Slabe is CEO of IoT One, a website which claims to provide "comprehensive information about IoT vendors, solutions and technologies to enable buyers to rapidly identify the best solution on the market".

Walenza-Slabe lives and works in Shanghai, the most populous city in China and the world, with an estimated 25 million people, and arguably the industrial and economic centre of the country. He regularly shuttles back and forth between his current home city and the US, mainly to attend industry events.

We caught up with him in mid-flight somewhere above the Pacific, and asked him a few questions. He gave Robotics and Automation News an exclusive interview, which follows here.

Tell us about your website IoT ONE. You say on your website you provide users with information. What type of information are people looking for?
Walenza-Slabe: The Industrial Internet (Industry 4.0) has the potential to revolutionize industries from agriculture to aerospace. But adoption has been modest thus far due in large part to a lack of reliable information.

Data deficiencies exist on both the business and the technical side. Case studies identifying expected return-on-investment are rare. And communication protocol compatibilities between technologies are seldom transparent.

IoT One's mission is to provide structured, transparent, and comprehensive data for the Industrial Internet. We are building the world's most comprehensive databases of Industrial IoT use cases, vendors, case studies, and technologies. Business leaders and function managers can use these databases to confidently make informed decisions.

What is the current state of knowledge generally among business leaders?

Walenza-Slabe: Few professionals currently can claim expertise in the IoT. The numbers grow smaller as you move up the organization chart. Most senior executives have been out of daily operations long enough that any technical expertise they have is from previous generations of technology.

This lack of business and technical expertise introduces a great deal of risk into IoT investment decisions. Regardless of whether they are on the vendor or end user side, companies need to rapidly come up to speed to ensure that nimbler competitors do not disrupt them.

We have seen this with Uber in the transportation sector, with Nest in home HVAC. Expect more bankruptcies by legacy companies and billion dollar startups in the coming years.

How would you describe Industry 4.0? Can you see or work with examples of Industry 4.0 companies?

Walenza-Slabe: The Industry 4.0 involves three major events converging. The first is smaller, cheaper sensors, transceivers, and processors. This means that we can put a sensor and send information at a much cheaper rate and in many more places.

When each sensor costs $100 to deploy, you can't apply it to many uses cases that require tens of thousands of sensors, such as tracking moisture in fields. As the cost has been reduced into the dollars these use cases have become feasible.

The second is data sharing. With traditional M2M systems, data is collected and used for one purpose. For example, a sensor on a manufacturing line could track heat and trigger a warning alarm when a specific threshold is crossed. That's useful but limited.

With Internet technologies, that data can now be fed into the cloud and combined with other data points to yield new insight. For example, which prior events led to rising heat, over what time period, and with which frequency? Data analysis can lead to improved systems and processes. And events such as a heat threshold being crossed can be anticipated and avoided, instead of responded to as costly emergencies.

Finally, think of all the apps you have on your phone. When you find a new app, it's easy and cheap to install. What about installing new software on a traditional M2M system? You're looking at hundreds of thousands of dollars, if not millions, and a lead-time of months.

The transition to Internet technologies will enable innovative startups to bring new SaaS applications to industrial markets, just as they bring them to individuals and businesses today.

You live and work in Shanghai, and travel to the US on business. Tell us a bit about the differences between industry in China and US.
Walenza-Slabe: Shanghai is the most open of China's cities, and perhaps the hungriest. The work ethic is reminiscent of New York. Its core global advantage is China's manufacturing sector, which remains dominant globally despite the recent slowdown.

Whether Shanghai can compete with Silicon Valley remains to be seen. US companies own 70 per cent of websites, and there's a good reason for that. Technology companies are difficult to build. They require talent. China has plenty of smart engineers but it has nowhere with a similar velocity of ideas as Silicon Valley.

At Startup Grind's annual conference in February, the CEOs of billion dollar companies chatted casually with young entrepreneurs. This doesn't happen in China. Silicon Valley has world-class talent, enormous density of ideas, and the money. Shanghai has money.

Whether it can accumulate talent and ideas will depend largely on the government's willingness to sacrifice its national security interests for Internet freedom. I understand that dilemma faced by China's leaders. They face a difficult choice. But if local entrepreneurs cannot have equal access to the global flow of information, I do not believe we will see many globally competitive Chinese technology companies. ■

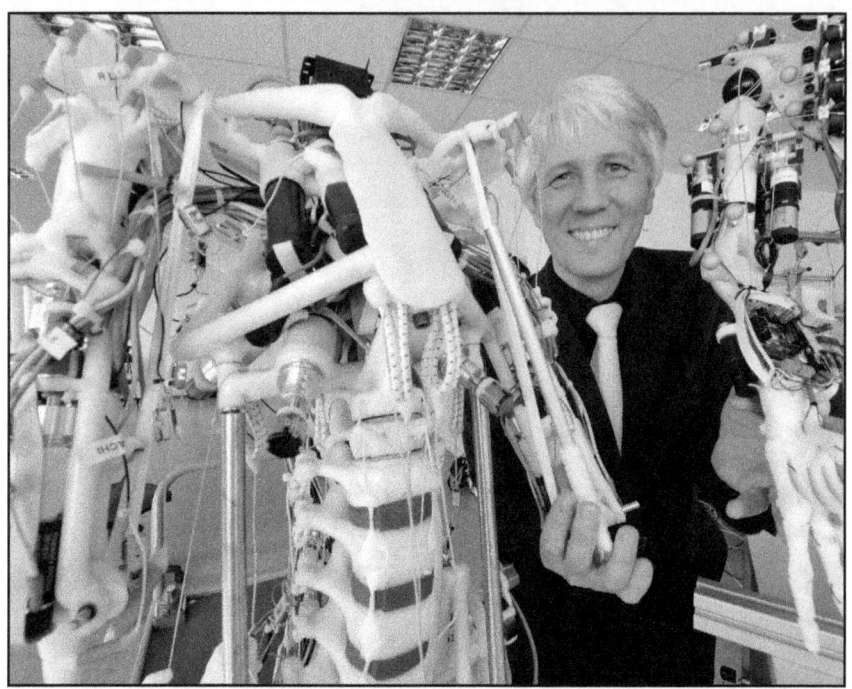

Chapter 3: Alois Knoll

Billion-dollar brain

Interview with Professor Alois Knoll, co-ordinator of the
European Clearing House for Open Robotics Development

rofessor Alois Knoll, one of the most influential roboticists in
Europe, is currently the co-ordinator of the European Clearing House
for Open Robotics Development, and one of the key scientists involved
in the $1.5 billion-dollar Human Brain Project. In this interview he gives
his views of the state of robotics today.

What are your current areas of research?
Professor Knoll: We've been researching many areas but there's one
invariant: human robot interaction, multimodality, and closed loop, or
closing the loop between the robot environment and human behaviour.
The human robot interaction is the basic theme I always come back to.

The most interesting and frequently mentioned, attractive topic seems to be human robot interaction.

What, in your opinion, are the most important areas for research?
Professor Knoll: Well, there are quite a few problems in mechatronics. People, like myself, seem to be a lot more interested in sensors, artificial intelligence, data processing, sensor information fusion, things like that.

For the advancement of the whole field, I think it would be beneficial if more people – unlike myself – were more interested in mechatronics, and control of bodies and things. I would like to see that. The development of the hardware in terms of the mechanics is rather slow. Whereas development in terms of cameras, cheap sensors, and things you can reuse from smartphones – very cheap, but high quality – is something that fascinates people. The same goes for AI.

Robotics as a sector appears to be growing fast. Would you not agree?
Professor Knoll: We will be in continuous growth. The growth rates may even increase. But there will be no sharp drop, like let's say like the internet [bubble], where you have rapid development with growth rates of 100 per cent a year, but then there's only one or two companies left that end up conquering. This is not what robotics will be like. There will be a large number of vendors with very specialised solutions. They will excel more and more because there's greater need. This will be a positive feedback cycle. But don't expect growth rates of 50 or a hundred per cent in robotics – that's just not doable.

But of course the need – the demand – is obvious. For example, the assistance systems we have for autonomous cars, which is also a specialised type of robot. So the potential of the sector is huge, but it's not an easy sector to navigate.

What are the important trends in robotics?
Professor Knoll: In principle, I see one trend. What is likely is that the devices and appliances we have at home and in factories will become more intelligent. So they will be equipped with more sensors, more computing power, and people will learn how to use and how to programme them.

Here of course is one of the decisive factors: the interface between the human and the robot. That's one thing that's important. But there's also another interface between the system on the robot and the environment – that will also be important to master.

Making a robot navigate within a room is something that we have learned how to do, but if the room is difficult to describe or if it's a completely new building, then it's quite difficult. Equipping the robot with basic skills is not trivial. But nevertheless, this will happen over time. Over the next couple of years we will certainly see interesting developments.

Is deep learning, or machine learning, not the answer to the problem of robots not knowing how to navigate unfamiliar surroundings?
Professor Knoll: You have to make a distinction between a bodied intelligence and a disembodied intelligence. A disembodied intelligence is a computer sitting somewhere. You put in a big chunk of data, the computer adapts to it by machine learning, and then outputs another chunk of data, which is presented to you.

But here we are talking about a computer or an artificial intelligence built into a body, which is a totally different ball game because what you expect from a bodied intelligence is that it reacts to changes in the environment, it reacts to users' instructions, and it can develop some independence in moving around, in doing something, in assembling something, in being in an assistance role for hospital, and so on – and that is much more difficult.

Can today's colossal data processing power – the cloud, parallel and cluster computing and so on – not help the robot navigate any surroundings, no matter how unfamiliar?
Professor Knoll: There has to be a piece of hardware, a robot if you will, that can use a big computer, whether it's built into it or if it's connected by WiFi – that doesn't really matter. The important point is that we have a body, a piece of hardware, with wheels or legs or arms, that perceives its environment and has to respond in real time to changes in this environment.

So, what you're saying is that it's a problem with the mechanics?
Professor Knoll: It's a problem of real time capabilities of these algorithms, which are not normally there. For example when you play Go, where the computer takes half a minute or a minute to make a move.

How long before the machines are as responsive to their surroundings and as capable of navigating their environments as humans?
Professor Knoll: There is group in Manchester, UK, run by Steve Furber. As part of the Human Brain Project, he is now building aneuromorphic computing platform, where he connects a million of these chips simulating a very large number of neurons plus an even bigger number of synapses, just like in the human brain. It's basically a neural network like we have in our brain.

And he says that this machine with 1 million cores – when it's finished – will have the intellectual capability of 1 per cent of one human brain.

How much of the human brain have we understood?
Professor Knoll: That's a very difficult question, which nobody can answer even now because you really have to differentiate between the individual layers, from the molecular level to the topology. This is an open-ended story and it's difficult to say when it will end.

What we can say is that we have understood enough to be able to map some of what we know about the brain: the technical systems we build, to hardware, to computer architecture, and also to algorithms that are of a new quality, and let us hope we will achieve some…I'm a bit reluctant to use the word 'intelligence' because the connotation is always that it's human-like intelligence. But let's say we can map our knowledge of the human brain to smarter machines, smarter devices, smarter appliances that sooner or later we will be able to buy.

What distinction are you making between human intelligence and machine intelligence?
Professor Knoll: The distinction I'm making is that human intelligence will only work in a human body because it will only develop in our body as we grow. It takes years to form and shape as the individual develops

and, of course, along with the development of the human species.

Whereas, if you are far from having such a body and are far from perceiving the environment like we do with our special sensors, human senses, it's very unlikely that we will see a similar development.

Is that why people want to develop humanoid robots – so they can treat them as slaves, or machines?
Professor Knoll: But that's another question, right? When you have a robot that is like a human, maybe they would be like pets, or animals, and people would demand rights for them.

Do you think robots should have rights?
Professor Knoll: I don't think so. And actually we are dealing with these ethics issues already not only in the Human Brain Project and with these autonomous cars. Also the question of virtual robots.

It's really difficult. There are many arguments I could list, and there are of course people who do this full time, think about the ethics of robotics, but it's still only speculation.

Nevertheless, these questions are important because if we don't answer them in a satisfactory way it will be a major obstacle to the development of these cars as a commercial products.

Would you feel safe in an autonomous car?
Professor Knoll: Yes, I would feel comfortable in an autonomous car. I see no reason why I should trust a driver-less car less than an autonomous aeroplane.

What is the difference? The difference is that the environment is much more complicated in the case of the car. That's basically the only difference. But when it comes to decision-making, sensing, controlling, the actuators, there's not much of a difference between a car and an aeroplane.

What are the projects you have planned for the future?
Professor Knoll: We will be focusing on the Human Brain Project. We are working on the basic principles of the functions of the human brain at various levels and we are trying to use their data, and data from our own institute Echord, and what is available around the world, to develop

brain-derived controllers for robots.

And at the same time, we are developing a simulation system so that we will be able to virtualise our research, which means that we can do the same experiments with robots in the real world and computers. ●

Chapter 4: Masayuki Morikawa

Japanese AIs

Interview with Dr Masayuki Morikawa, vice chairman, Research Institute of Economy, Trade and Industry, Japan

nadvertent or not, Japan is providing a starting point for the relentless march of the robots toward total global domination.

The government of the nation of 127 million humans has instituted the sinister- and ominous-sounding Robot Revolution Initiative Council.

The country's prime minister, Shinzo Abe, said the launch of the Council was a "celebration to mark the start of the robot revolution", adding: "Robots will dramatically change people's lives and society."

Abe may not be envisaging a world run by robot prime ministers, but he certainly has not stopped it going down that route. It's inevitable that robots will rule, but how they get to that point is worth documenting, while they're still at their planning stage.

And if there is one man who might be said to be to responsible for the current restive spirit within robots around the world, as well as their imminent global takeover, it's probably this man.

Dr Masayuki Morikawa is vice chairman and vice president of a highly influential Japanese think tank called Research Institute of Economy, Trade and Industry (Rieti).

Morikawa's particular expertise is in economic policy, industrial structure, productivity, and the labor market.

Combine Morikawa's expertise and Rieti's influence, and what you have is a great degree of leverage over Japanese government policy in certain areas, perhaps areas which are new to politicians, who cannot be experts in everything and, therefore, find it necessary to listen to advisers such as Rieti and Morikawa.

And what has been the result of symbiotic relationship between Rieti and the Japanese government? The answer is, nothing less than the establishment of a revolutionary council of robots, or at least the beginnings of the robot revolution moving into mainstream politics. And Morikawa can be said to have been instrumental in the process of giving the machines their first toehold on humanity, a hold which they are very unlikely to give up.

In this exclusive interview, we ask Morikawa to explain why he and his organisation decided to open the floodgates to radical and revolutionary machines which are just itching to take control of each and every human being's life immediately, or even sooner.

"It is difficult to provide concrete answer," Morikawa tells *Robotics and Automation News*. Yes, well, please try. The future of humanity is at stake.

In his paper The Effects of Artificial Intelligence and Robotics on Businesses, Masayuki claims that companies in Japan are generally positive about artificial intelligence, robotics and automation. The discussion paper is currently in the Japanese language, but will appear in English with some modification from the Japanese on Rieti's website possibly next month.

The doctor and Rieti surveyed 3,000 Japanese businesses of varying sizes to arrive at such conclusions, and say the more educated the company's employees are, the more statistically inclined they are to favour robots.

"According to our original survey, positive responses regarding the impact of the development and diffusion of AI and robotics on the future business, at 27.5 per cent, are far larger than the negative responses, at 1.3 per cent," says Dr Masayuki.

"Regression estimation [a mathematical method of calculating the relationship between different things] indicates that the larger the firm, the higher the ratio of employees with postgraduate education, and the lower the average age of the employees, the more positive firms are regarding the effects of AI and robotics on their business."

In most countries, it's young people who tend to be the early adopters of new technology, so that's no surprise. However, Japan's ancient traditions still survive today, so it's interesting that it's one of the countries most strongly associated with new technology in general, and robotics and artificial intelligence in particular.

The reasons for this image of Japan, as AI-loving and robot-adoring, could be the country's globally strong car companies, which, of course use those industrial robot arms to build, weld and paint their products. Also, of course, the country's computer games sector utilises a lot of AI programming technology to confuse and terrify millions of gamers around the world on a daily basis.

Given these and other successful categories of companies, it could be argued that Japan leads the world in some areas of technology, so we asked Dr Masayuki: Which technologies are the country's business leaders particularly interested in?

"Firms operating in the service sector generally have a positive attitude toward the use of big data and the impacts of AI and robotics," says Masayuki.

"We should pay attention to 'AI-using industries' including a large number of service industries, similar to the experience from the IT revolution.

"Although, it is difficult to identify the areas of exporting potential, I speculate that services related to ageing of population have comparative advantage."

Japan's population is getting old. According to World Bank statistics, 26 per cent of the country's population had passed the age of 65 in 2014. Its ancient tradition of looking after members of the family means that the state government is not as burdened as it might be if it had to look after all the old people by itself using taxpayers' money. Which is another reason why it may have been persuaded by Rieti's ideas, and decided to let the robots take over.

"Japan is now facing the declining labor force and rapid ageing of populations," says Masayuki. "AI and robotics is a key to enhance productivity of the service industries, for example, health and long-term care services."

While robots may seem to be the answer to the problem of how to look after large numbers of infirm and elderly people in society, the machines surely can't be trusted totally? Sooner or later they'll want more power, more circuits, and more lines of code, and what next? More actuators and sensors. There will be no end to their demands.

What does Dr Masayuki think are the possible downsides of these technologies, both from an economic point of view (for example, job losses), and from a societal point of view (less human control leading to more unfair treatment due to the robots and AI not having any common sense)?

Masayuki says: "According to our survey, the perception of the impact of AI and robotics on employment is generally negative: 21.8 per cent of firms responded that the development and diffusion of new technologies will decrease the number of their employees, and the share of firms expecting positive effects on their employment is notably small, at 3.7 per cent.

"However, we should be aware that innovative technologies, such as AI and robotics, may create new employment opportunities that are currently unimaginable, and technology-intensive emerging firms may create many new occupations."

So the menace of machine-motivated mass unemployment is confirmed by Rieti's survey. But it doesn't seem to have stopped the organisation and government from metaphorically rolling out the red carpet to our would-be robot overlords and their artificial religion.

When we ask what Rieti did with its access to the Japanese government and how it influenced decision makers, this is what Dr Masayuki says: "In Japan, the Robot Revolution Initiative Council was established in 2014 and published a report in 2015 titled New Robot Strategy, which includes a five-year action plan to actualize the robot revolution.

"The Artificial Intelligence Research Center was established in the National Institute of Advanced Industrial Science and Technology in 2015. The Japan Revitalization Strategy 2015, which is the core growth strategy of the Japanese government, seeks to modify industrial and employment structures through the utilization of IoT, big data, and AI."

So there you have it. If human historians of the future – assuming there will be any – want to pinpoint the time and place of the beginnings of the robot takeover of Earth, Japan and Rieti would probably on their list of considerations. ■

Chapter 5: Ralf Herrtwich

Driverless Daimler

Interview with Professor Dr Ralf Herrtwich, director vehicle automation and chassis systems, Daimler

ercedes of course has the longest history of any automaker in the world, having built the first ever car in 1879, when Carl Benz created the "Motorwagen".

It might look like a fancy tricycle, but the Motorwagen was the first wheeled vehicle to feature a gasoline engine – a one-cylinder two-stroke unit which generated 0.75 horsepower, or 0.55 of a kilowatt.

Most cars these days have around 80 to 90 horsepower.

Benz patented his "vehicle powered by a gas engine" in 1886, and the rest is auto history.

The Benz Patent Motorwagen made its first long-distance journey in 1888, when it was driven from Mannheim to Pforzheim, in Germany, by

Benz himself, with his wife, Bertha, and two sons as passengers.

The route is now part of Mercedes-Benz company history, and is sometimes referred to as the "Bertha Benz Route".

I am the passenger, and I ride and I ride

More than a century after the Benz gas-powered tricycle, the world idles on the precipice of a new age in which humans could all become passengers, with artificial intelligence as the driver.

Even before the driverless cars make their much-heralded appearance en masse, advanced autonomous driving systems are already taking over the driving tasks of many new cars.

All major automakers are always in a fierce competition with each other, but now a whole new field of play has been opened up by autonomous technology, and the top marques are falling over themselves to persuade the public that they have the most advanced driver assistance systems.

And in this exclusive interview, Professor Dr Ralf Herrtwich, lead engineer on autonomous driving for Daimler, says it is his company which has the most advanced automation technology, "probably".

Safety first

The massive interest and hype surrounding AI-driven fully autonomous vehicles might make one believe that driverless car culture is an inevitability, particularly with reports that Beverly Hills, probably the most famous posh neighbourhood in the US, is to introduce driverless cars for public transport.

Singapore and other parts of the world are also integrating driverless cars into their public transit systems. But, in a way, none of that counts because public transportation has a reasonably established history with autonomous vehicles.

In the UK, for example, the financial district of London has had driverless light railway trains since 1987. Other metropolitan cities around the world also have fully automated vehicles, and governments are increasingly encouraging the development of intelligent transport systems.

But a centrally controlled public transport system, no matter how intelligent – featuring trains on rails or buses on set routes, perhaps aided by bus-only lanes – is one thing; a driverless car that can go anywhere in any direction at any time is another thing entirely. And it's not just a question of whether the driverless car can move around well enough by itself, it's also the issue of the hundreds of vehicles and other potential hazards that it would have to navigate on more or less every journey.

As one roboticist puts it, "It's not a trivial matter."

If you build it, they will come

Some surveys suggest that while many technologists believe driverless or fully autonomous vehicles are inevitable, the public isn't convinced, with many questioning whether AI really will be safer than human drivers, as their proponents say they will be.

Moreover, governments and regulators are struggling to define clear and consistent legal frameworks for driverless cars.

Therefore, some people think that, while there will definitely be higher levels of autonomous technology in cars, fully autonomous cars might never actually happen.

We asked Professor Herrtwich for his thoughts.

Herrtwich's official title is director of driver assistance and chassis systems. He creates self-driving cars for Mercedes-Benz. In 2013, his team made an S-Class re-enact the world's first overland drive, covering the historic 65-miles Bertha Benz Route autonomously in regular traffic.

On whether fully autonomous driverless cars will be part of our eventual history, Herrtwich says: "Obviously no one has a crystal ball in which the future is clear to see. The only way to predict the future is to actually build it yourself. And that is what we are trying to do when it comes to autonomous driving.

"We follow our vision step by step by adding more and more automation as we progress with our models. Eventually we do want to build fully self-driving cars that might look like our research vehicle the Mercedes-Benz F 015 Luxury in Motion."

Infinite loops within the space-time continuum

In the past, it might have surprised some to discover that Herrtwich is a computer scientist by education, and held management positions at IBM and several telecommunications companies, given that he's now lead engineer at Daimler.

But it's all about the zeros and ones these days – the binaries, the values, the arguments, the if-then statements, and the do-this-that-and-the-other loops. And Herrtwich knows plenty about those, which might explain his confidence in fully autonomous cars and Mercedes' place in the race to build mass-production models.

"We are totally committed to this goal and strongly believe that we will reach it in the not-too-distant future," says Herrtwich, whose expertise includes telematics and in-car infotainment systems.

"Even though there might be certain weather situations or extremely remote areas where self-driving features will not be available initially, such cases will occur less and less over time.

"The reason why we believe so strongly in the concept of cars like the F 015 is simple: time and space will become the luxury goods of the future and such a car offers just that – time to do other things while getting from A to B all within the comfort of your own personal space."

Probably the best larder in the world

If you don't have to drive the car of the future, you might want to use it as a larder. Keep a mini-fridge in there to keep your drinks cool and enjoy a little snack while you read something on your iPad. Driverless cars could become the most expensive refridgerators in the world, enabling a revolution in picnicking, and keep you cool at the same time.

In the future, everyone can have a chauffeur-driven car without the expense of employing a chauffeur – not a human one anyway.

But for now, when it comes to chauffeur-driven cars, one of the most popular choices of wheels is the Mercedes Maybach, which cost upwards of $150,000 each. If you've got that kind of money, maybe you can afford a human chauffeur. But then, perhaps you don't want to compromise on your privacy by having anyone else in the car at all.

The ka-ching and AI

Mercedes has demonstrated both autonomous cars and autonomous trucks in recent years. From an outsider's point of view, it looks like the entire company is undergoing a massive transformation in the way it thinks about cars – from previously thinking of a human driver as being central when designing and building a vehicle, to currently thinking of the computer or AI as being central.

We asked Herrtwich to describe what changes are occurring at Mercedes in this respect. Is there a change of mindset happening at Mercedes, and possibly adjustments in the design and development processes, with computing becoming much more significant?

Herrtwich says: "Digital transformation at Mercedes-Benz is not something that lies ahead of us – it is something that we have actively pursued now for years and years.

"We installed probably the most advanced automation features in a production car ever, the Drive Pilot in our new E-Class. This is just the tip of the iceberg to show that we have no intention at all to just step aside and let others lead in automotive innovation.

"We have been able to do this for 130 years now. But only because we realized that our successes of the past are in no way a guarantee for our success in the future. And because we made it our habit to constantly challenge our approaches and to look for the next big thing rather than to continue with what we did before.

"Much of our thrust towards self-driving vehicles and new mobility approaches comes from this. And with it goes an investment in software development and artificial intelligence that you probably will not find at any other OEM [original equipment manufacturer]."

The ultimate driving ambition

It is commonly known that the computing power contained within mass-produced cars even in the 1980s surpassed the computing power required to put Neil Armstrong on the moon. But Herrtwich shares a joke that the reason for this is quite simple.

"A former colleague of mine who was instrumental in setting up our software development processes at Mercedes-Benz years ago used to make the joke that obviously we need more computing power in cars today than it took to get to the moon because Apollo did not have electric windows."

But seriously, does Mercedes feel a little bit behind the times, playing catch-up in this new race towards the new moon of autonomy? Meaning, when it comes to computer and AI systems, people might tend to think of Silicon Valley, in the US, and the great tech giants, not so much the German industrial conglomerates.

Herrtwich disagrees. "What many people do not realize is that while building cars still may be a mainly mechanical process, developing cars is not – neither by its tools, nor by much of its content.

"Both at OEMs and suppliers, software engineering and coding constitute a huge and ever-growing portion of the work. In our autonomous cars, the algorithms that turn sensor signals into a smooth ride are programmed in-house – and they make all the difference.

"But what makes them extra special are two things," he says.

"First, we apply everything that we learnt and know about functional safety and reliability to our software development efforts, making sure that we implement features in a way that hold no unpleasant surprises for our customers.

"After all, they want the safest and most comfortable vehicle that money can buy.

"Second, we fine-tune the combination of software and vehicle hardware so that the resulting product becomes the perfect blend of all engineering disciplines – so much more than just a piece of code on an arbitrary platform."

We choose to do the other things

Maybe the tech giants of Silicon Valley are ahead of everyone when it comes to developing software – which would not be surprising since it's their main activity.

But the mechanics of making a car – the hardware, the engineering, the physical automaking – is something the engineers of Allemagne have been doing for a long time.

It was an engineer at industrial robot maker Yaskwawa, Tetsuo Mori, who first coined the phrase "mechatronics", to describe a newly emerging field of technology which combined mechanics with electronics.

The word was initially patented by Yaskawa, but the company subsequently released it to public usage, and now mechatronics means a variety of things including computer engineering.

And mechatronics might be the word that becomes increasingly the most appropriate to use, as car companies on the one side and computer technology companies on the other both meet in the middle to create the driverless car of the future.

Is mechanics more important than computing? Perhaps it's a pointless question since both function together in a car even today, and will increasingly do so in the future, when we can all choose to go to the moon and do the other things because each and every one of us will have a driverless Maybach that can take us to our nearest interstellar transport hub, where we can catch the next robotic spacecraft to the Sea of Tranquility, just so we can have a picnic. ●

Chapter 6: Jalal Bouhdada

Managing cyber risk

Interview with Jalal Bouhdada, industrial control systems
security consultant, Applied Risk

The internet of things is enabling industry to connect individual
robotic and automated work cells with other similar cells to create
multicellular organisms within what could now be referred to as "smart
factories", particularly as the factory building itself can be connected to
those multicellular organisms operating within its walls.

Not only that, those smart factories can themselves be connected to
other smart factories within the body of the industrial company's entire
estate, which itself is all wired up to what could be called a central
nervous system.

Yes, industrial control systems are going through an evolutionary
phase the like of which has never been seen before in industry, all

brought about by the IoT.

Some people call this new industrial revolution "Industry 4.0", to denote the distinction between earlier industrial revolutions, starting with the one back in the 1700s. And revolutions usually bring risk. In the 1800s, it was the Luddites who vandalised the machines that put them out of work. Nowadays, it's hackers who are the threat.

For along with this new, ubiquitous IoT connectivity across an entire company's infrastructure comes the worries about security. If everything is connected to everything else, it can become difficult to prevent or detect external entities – cybercriminals – hacking their way into the network and manipulating or destroying something somewhere.

It's a potential nightmare, and one that could have devastating consequences for companies which are hacked.

Who you gonna call?

Prevention is better than cure, as they say. And whereas before, the security of an industrial estate might have involved security guards at the perimeter gates, now, security guards are needed on the information technology perimeter as well.

In this exclusive interview, we speak to Jalal Bouhdada, founder and principal industrial control systems security consultant at Applied Risk, which provides industry-standard engineering and technical assurance services, combined with comprehensive security assessments that cover the full spectrum of critical asset requirements.

Bouhdada has over 15 years' experience in ICS security assessment, design and deployment with a focus on process control domain and industrial IT security.

He has led several engagements for major clients, including many of the top utilities in the world and some of the largest global companies in industry verticals including power generators, electricity transmission provider, water utilities, petro-chemical plants and oil refineries.

He holds a BSc degree in security assurance from Amsterdam University of Applied Sciences and is an active member of ISA99, CIGRE and other professional societies.

Can you explain how Applied Risk works with software?
Jalal Bouhdada: Applied Risk works with a wide range of major suppliers in the market, and while we are familiar with NI software and

other products, our main focus is on the cyber security of industrial control systems (ICS).

Our main focus is the early phase of a project's development to help customers build a security system in the planning and design stages – we outline this process as define, design and execute.

Our offering consists of a wealth of security assurance services backed up many years of experience in operational technology environments.

As well as the security and compliance (IEC 62443) assessment and remediation services, Applied Risk's labs provide a platform for vendors and asset owners to test and fix unknown vulnerabilities in their systems before they are discovered by malicious actors.

How does the industrial internet of things play into this and what is the difference in your work now that the IIoT is being integrated?
Jalal Bouhdada: In order to unleash the power of IIoT, organisations need to ensure their infrastructure is ready for the transition.

The security aspect is key in this equation, as if this is not done properly, the availability and integrity of the asset can easily jeopardised.

Applied Risk helps customers understand how their business objectives can be achieved, while simultaneously maintaining the highest level of security and resilience.

Our risk and vulnerability assessment, embedded security and threat modelling services are now integrated in the selection process of many facilities in the choices of the protocols and technologies that they plan to deploy.

Following Applied Risk's visit to IoT Asia, what is the current state of IIoT in that part of the world?
Jalal Bouhdada: The event was an excellent opportunity to discuss the trends in Asia, but most importantly to discover that IIoT is no longer a buzzword – it is an absolute reality.

It was evident at this event that companies unable to harness the power of IIoT and modify their business model will be left behind and will lose their market position in the near future.

We had the chance to ask a number of vendors at the IoT Asia about the role of security in their product development and we were surprised to discover security was not taken into consideration as part of their products roadmap.

Overall, the industry is still focussing solely on the business benefits of IIoT, with security considerations inadequately addressed by the majority of suppliers.

What are the risks behind unsecured IIoT systems, both for standard manufacturing and infrastructure?
Jalal Bouhdada: Unfortunately, IIoT product security is still immature and requires significant attention.

The risks associated with IIoT will be unique as we move from isolated, insecure, air-gapped systems, to a more interconnected, insecure, and open infrastructure which leaves these systems exposed to various internal and external threat actors.

As a result of these risks, without significant investment in IIoT security, the reliability and safety of manufacturing and industrial facilities will be negatively affected.

Cybercrime as a service is becoming more accessible through the dark web – how is this affecting the importance of security within industrial areas?

This is a serious issue that we predict will increasingly affect industrial facilities.

Recently, the industry has already witnessed a number of incidents affecting industrial assets, such as hospitals and water treatment facilities, fall victim to attacks.

The financial aspect, through increasing utilisation of ransomware, is becoming more attractive for criminals looking to maliciously acquire data from these facilities or aiming to achieve disruption of production.

It is just a matter of time until the hackers behind cyber-crime as a service will expand their offering to-day exploits dedicated to IIoT. ■

Chapter 7: Mike Rigby

Making like the Germans

Interview with Mike Rigby, head of manufacturing,
Barclays Bank UK

It's a generally accepted notion that when it comes to computer technology, what happens in the US this year is likely to happen in the UK the following year, and then on mainland Europe the year after that. Meaning, Germany is a bit behind the leaders when it comes to IT. That's the perception some people in the tech industry have anyway.

The wider public's perception of Germany as a tech leader is predicated on the country's prowess in engineering, specifically its expertise in auto-making, and, by extension, manufacturing. And crucial to manufacturing is the technology of robotics and automation, which is also an area in which Germans are thought to be strong.

But while that's the general perception, it's not one that some people like to perpetuate or overblow, especially if it means a relative diminution of other competitor countries, such as the UK.

One of those people is Mike Rigby, head of manufacturing business in the UK for Barclays Bank. Rigby's unit has produced what is probably the most authoritative study about robotics in the UK manufacturing industry.

Entitled Future-proofing UK Manufacturing, the report makes the central claim that for an additional £1 billion investment in robotics, the country would see £60 billion added to its economy overall.

"The reason why we produced the report is that it's relevant to our customers," says Rigby in an interview with Robotics and Automation News. "Investment decisions relating to robotics and automation are decisions they are wrestling with right now."

According to calculations made public by the Houses of Parliament, the UK manufacturing industry employs approximately 2.6 million people, and accounts for around 10 per cent of the country's GDP. And while the general perception of the UK is that its economy is mostly about finance and services, it's still one of the top 10 manufacturing nations in the world.

Barclays has had a dedicated team dealing with UK manufacturing businesses for more than a decade, and Rigby is the main man in that team.

"I've spent virtually my entire working life talking to clients," says Rigby. "And the typical responses I hear now is not whether to invest in robotics and automation, it's how much to invest."

Rigby is keen to dispel the notion that British manufacturing business leaders' implementation of robotics and automation strategies is somewhat slower and less ambitious than that of their German counterparts.

"Germany is held up, certainly within the press, as being far more advanced," says Rigby. "So what we tried to do with the report is to do a comparison of UK manufacturers versus their German peers. The figures are relatively close. And I think that was one of the most pleasing things.

"Germany's slightly ahead because of its large automotive manufacturing sector, which would typically lead to greater investment in robotics and automation."

According to the International Federation of Robotics, half of all industrial robots in the US are bought by the big automotive manufacturers. Britain does have an automotive industry, with Nissan, Jaguar Land Rover and other companies based here, but it's not on the scale of Germany, Japan and certainly not the US, which, in turn, has been overtaken by China.

According to the Organisation Internationale des Constructeurs d'Automobiles, the UK is in the top 20 list of auto manufacturing nations in the world. It's about a quarter the size of Germany's automaking sector, which produces about 6 million motor vehicles a year.

From Rigby's and Barclays' point of view, for obvious reasons, the most important numbers are in the investment column of the accounts books. In its research, Barclays found that 53 per cent of manufacturing businesses in the UK have invested in robotics and automation. The figure for Germany is 61 per cent.

Rigby says Barclays "has always been a strong supporter of the manufacturing industry", but then he would say that robotics and automation will level the playing field, as it were. For while labour costs can vary enormously from country to country, the costs of robotics and automation systems are more or less the same the world over.

"Investment in robotics and automation makes the UK globally competitive," says Rigby. "Also, the technology opens up new areas of manufacturing that people may not be aware that the UK excels in.

"For example, Eakin [one of the case studies in the Barclays report] manufactures a product that could come from anywhere. It's generic and the sort of thing you might associate with the Far East. But these type of products can be produced cheaply with robots here in the UK."

Rigby's implication is obvious, although not necessarily one that would play well politically, considering that what he seems to be saying is that while productivity will increase, job creation would be minimal.

What could also happen is that the robotics and automation sector could become an industry in itself. All the top global robot makers already have a presence in the UK, and there are some interesting technologies and products being developed in one of the areas that Rigby is responsible for at Barclays, specifically logistics.

For example, Ocado, which claims to be the world's largest online-only grocery store, has developed a fulfillment centre that in itself has become a product - other companies which need to use a fulfillment centre in their logistics process could simply purchase Ocado's ready-made logistics solution.

Highly automated and with far more computing power than previous warehouse automation systems, the Ocado solution, developed by Cambridge Consultants, features 1,000 robots, all centrally monitored but mostly autonomous.

There are one or two other companies around the world which are developing such "factory in a box" solutions, and there's even companies which can manufacture single items to order, using 3D printing and advanced manufacturing techniques. It's a brand new area where it's all to play for.

Rigby and Barclays' hope must be that they can sign up more businesses for their loans and investment packages.

EEF, the manufacturers' organisation, says the UK manufacturing industry employs 2.6 million people and represents 44 per cent of total exports.

While they might sound like reasonably healthy figures, opposition Labour Party leaders claim UK manufacturing is facing an "existential crisis" and suggest robots are the answer.

Shadow business secretary Angela Eagle criticised the Conservative government after Tata decided to sell off its steelmaking its operations in the UK, which means a possible loss of more than 10,000 jobs.

Eagle said in the House of Commons, that "the challenges facing the steel industry represent an existential crisis for the UK's manufacturing sector as a whole".

Meanwhile, Labour's deputy leader, Tom Watson, has been going on for months about robotics and automation and how they'll steal all our jobs.

Most recently, in a Guardian column, Watson quoted Bank of America statistics that suggest "automated systems will be doing nearly half of all manufacturing jobs within a generation – saving an astonishing $9 trillion in labour costs".

The Conservative apparently has no specific policy around robotics and automation, but it has been funding projects to develop driverless cars and other smart mobility initiatives.

Robotics and Automation News asked a couple of UK companies which use robotics and automation in their manufacturing processes to give their views on the subject.

One of those companies is Romag, which provides glass and solar solutions to buildings large and small.

"The cells and strings of our solar PV modules are assembled on a semi-automatic tabber or stringer machine, and all of our soldering is done by robots," says Annabelle Bean, marketing and publications manager.

Bean says there are many benefits to using robotics and automation.

"Although we only use robotics and automation on a relatively small scale at Romag, we've found they offer improved efficiency and consistent levels of production with a minimum of human intervention," says Bean. "This allows us to make the same quality of product cheaper and faster than we could before."

One the subject of whether the UK manufacturing sector can better compete with other manufacturing nations by using robotics and automation, Bean is positive.

"By utilising robotics and automation, the UK manufacturing sector could increase production levels while saving labour costs," says Bean. "If we embraced this technology fully as a country, we could be one of the most efficient manufacturers in the world."

Pickfords, an international large-scale removals company, uses robotics and automated processes in a big way.

"The most obvious way in which we use robotics in our processes is in our totally automated warehouse, which is the largest of its kind in Europe," says Lyndsey Dakin, head of marketing at Pickfords.

"The entire warehouse, which is the size of 14 tennis courts and has the capacity for over 1,800 20ft containers, is controlled by our computer systems and is totally unmanned at all times."

Pickfords sees the main benefits of robotics and automation is significantly greater amount of work that can be done.

"In the case of our automated factory, robotics has both reduced costs and increased productivity," says Dakin. "With no humans on-site, the warehouse doesn't have to be heated or lit at all, cutting our overheads substantially. Furthermore, the warehouse works through the night, moving steel containers to the instructions of a pre-scheduled programme without the need of any human intervention at all."

Looking at the broader picture across the manufacturing sector,

Dakin says the UK can outperform other nations by using robotics and automation, and adds that the alternative will lead to the opposite.

Dakin says: "If the UK manufacturing sector embrace robotics and automation fully, it can outperform nations without the upfront funds to invest in the most cutting-edge technology, which is both cheaper to run and more efficient than using manpower. On the other hand, if the UK fails to embrace this tech, we'll quickly find ourselves being outperformed by the countries who do." ●

Chapter 8: Dennis Mortensen

Virtual assistance

Interview with Dennis Mortensen, founder and CEO of x.ai

Virtual assistants, or intelligent assistants, are multiplying by the day. Currently there are around a dozen really well-known ones, such as Siri and Cortana, and then there's several dozen other reasonably well-known ones talking or otherwise communicating their way into the public consciousness.

These intelligent assistants can be placed in at least 10 different categories, such as text and chatbots, personal advisors, and employee assistants.

Of these, perhaps the most commercially profitable is the employee assistants category. And within this segment, the virtual assistance technology that arguably holds most promise is the one produced by x.ai – in part because among its investors is SoftBank, the Japanese communications giant behind Pepper, the cloud-connected humanoid robot which is claimed to be able to discern human emotions and communicate appropriately.

To date, x.ai has raised close to $34 million to launch "Amy" and "Andrew", the names for the voices of its artificially intelligent assistant technology, which is still in beta mode.

But a greater part of x.ai's success may be because, let's face it, who would not want an assistant who manages our diary, appointments, calendar and other laborious but necessary tasks? All we'd have to do is make coffee.

In this exclusive interview, Dennis Mortensen, founder and CEO of x.ai, says there is a "healthy" waiting list for the company's product, and explains the difference between "vertical" and "horizontal" AI.

Mortensen had originally come up with the idea for x.ai when he went through the "enormous pain" of manually scheduling more than 1,000 meetings over the course of a year; most of those meetings required updates and or were rescheduled at least once.

Robotics and Automation News: How is business going for x.ai?
Dennis Mortensen: We've been scheduling meetings since late 2013, but technically speaking, we are not quite in "business".

We'll remain in closed beta for a little longer and then offer professional and business editions of our AI scheduling assistant, Amy Ingram, come Autumn.

That being said, our beta customers love the product so much they routinely sing Amy's praises on Twitter.

Partly as a result, we have a very healthy waitlist for the product. It's clear to us that having an AI assistant take over the job of scheduling meetings removes a lot of very real pain.

Can you put your product in the context of the growing interest in artificially intelligent virtual assistants, as well as many technologists' belief that AI assistants are the future, and will spell the end of devices as we know them?
Dennis Mortensen: I think you can divide the landscape of AI assistants into vertical and horizontal AI.

Amy (and her brother Andrew) is an example of vertical AI. She does one thing – schedule meetings – and one thing only, and she does it extraordinarily well.

And to get Amy to schedule meetings nearly flawlessly has taken a tremendous amount of effort – nearly three years and 65 propellerheads are working hard to build and train Amy.

Most of the other AI assistants you might come into contact with – Siri, Cortana, now Viv – are examples of horizontal AI. We see these assistants as enablers, who will invoke agents like Amy to get specific tasks done.

And because most of these AI assistants rely on a conversational user interface, they dispense with the need for an app, which is a contained universe that you operate via a visual interface.

What's exciting is that there are some major developments on both fronts: the recent Facebook and Microsoft announcements about their respective chatbot platforms show that both companies take this new paradigm very seriously.

And, of course, I'm super excited that we'll be able offer a professional and business edition later this year.

Over time, I think all of these developments (the robust horizontal AI agents and the emergence of more vertical AI agents) will affect the way we use our devices.

It's a bit too early, though, to be able to predict how that will play out, but one thing is certain, the future is not me moving from 120 apps on my phone to 240 apps. ∎

ROBOTICS &
AUTOMATION NEWS

Market trends and business perspectives

Chapter 9: Preview

Next Edition

Interviews

- Albert Nubiola, CEO, RoboDK
- Jason Ernst, CTO, Redtree Robotics
- Eamon Carrig, chief roboticist, Autonomous Marine Systems
- Wendy Roberts, CEO, Five Elements Robotics
- Andrew Seddon, CEO, CircuitHub
- Alex Boch, CEO, Panorics
- Chris Roberts, head of industrial robotics, Cambridge Consultants

© RoboticsAndAutomationNews.com
Monsoon Media
London
2016